THE STORY OF THE ICE AGE

Rose Wyler and Gerald Ames

Illustrations by Thomas W. Voter

SCHOLASTIC INC.
New York Toronto London Auckland Sydney

Cover photo of Hubbard Glacier Alaska,
Copyright A. Keler/Sygma.

ISBN 0-590-41446-1

 Published by Scholastic Inc., 730 Broadway, New York, NY 10003, by arrangement with Harper and Row, Publishers, Inc.

12 11 10 9 8 7 6 5 4 1 2 3/9

Printed in the U.S.A. 40

Contents

SEVERAL times in the past, the lands of the earth were attacked by a mighty enemy: ice. Moving sheets of it buried the Antarctic Continent and Greenland. Other sheets overran half of Europe and North America. Blue ice cliffs plowed over valleys and plains and mountains. Only the tallest peaks rose above the ice, like islands in a white sea.

Those were dangerous times, yet bands of people in Europe and Asia dared to roam the borders of the ice, hunting the reindeer, the woolly rhinoceros, and the mammoth. Several kinds of great beasts died off during the Age of Ice, but the people lived.

The hunters had no way of writing and left no account of their adventures. Until about a hundred years ago, no one in the world knew they had ever existed or that there had ever been an Ice Age. Then bones of unknown animals and people were dug up, and hints of mysterious events were found. Scientists fitted the clues together, bit by bit, until they unfolded a strange chapter in the history of the earth.

A Mysterious Monster

IN the year 1799, news from arctic Siberia startled the world. A monster had been discovered there. According to reports, a fisherman had seen his dog run to a bed of frozen gravel and begin eating something. The man went to investigate and found a huge body, partly thawed, sticking out of the ground. The monster resembled an elephant, but had a coat of wool.

Here was a riddle for scientists. They said, "Elephants belong in the hot lands of Africa and India. How did they ever get to the Arctic?" Then they thought of their history books and remembered that in ancient times an army from Africa had brought elephants into Europe.

The commanding general, Hannibal, used them as living tanks. He terrified the Romans by marching his beasts over the passes of the Alps mountains into Italy.

"Did some of Hannibal's elephants escape?" people asked. "Did they stray across Europe and travel 4,000 miles to arctic Siberia?"

Although several more such dead monsters were found, no one ever saw one of them alive. The Siberians thought they lived underground and died if they came out into the sunlight. So they named the beast *mammut*, which in their language meant "earth creature." We use this word, but we spell it "mammoth."

Scientist-Detective

From time to time other mammoths were found. Their bones and hides were sent to museums, where scientists studied them.

One of these scientists was a Swiss professor, Louis Agassiz.[1] Ever since boyhood he had loved to explore nature. He had learned to dive underwater and catch fishes with his bare hands. He collected fishes as some people collect butterflies.

When he was a young man, Louis went to a German university to study animals of the past and present. "How different the ancient creatures were," he said. "I wonder why so many of them have disappeared. Did some disaster wipe them out?"

One day Louis visited the natural history museum in the city of Stuttgart. There he saw mammoth bones and a piece of hide that had been sent from Siberia. The hide was still covered with reddish wool, from which long black hairs stuck out.

Louis stared at the bones and the hide. He imagined he could see the wasteland of ice from which they had come.

"I wonder," he said to himself, "if any kind of elephant, even a woolly one, could live in a place as cold as the Arctic. How did it ever get there? How did it die, and how was its flesh preserved for ages?"

None of his professors could answer these questions. But Louis did not stop thinking about them. When he himself became a university professor, the image of the woolly mammoth often came to his mind. He could not forget the riddle of the vanished creatures.

Years later Agassiz was to find a key to the mystery. But he discovered it far from Siberia, in his own country, Switzerland.

[1]Say: AH-gah-see.

Ice That Moves

Agassiz had heard some curious things from a friend of his, an engineer named Charpentier.[1] This man spent his vacations exploring in the Alps. There he noticed something very strange. Many of the boulders lying about in the lower parts of the valleys were entirely different from the bedrock of the neighborhood. Charpentier decided to find out where the boulders came from. He hiked up the Rhône Valley, searching for bedrock like the boulders. At last he found it, high up in the mountains. The boulders must have come from there.

"What moved the boulders?" Agassiz asked. "Flood water?"

"No," said Charpentier. "That's what most people think, but the peasants who live in the mountains say the boulders were carried by glaciers."

"But glaciers are high in the mountains. And ice doesn't move."

"Glacier ice really does move, like a slow stream," said Charpentier. "The peasants have watched boulders lying on the glaciers, and noticed that they move downward a certain distance every year. They think the glaciers once were much longer than now and carried chunks of rock from the summits all the way down to the lower valleys and the plain. They call the boulders 'foundlings' — homeless children."

Agassiz could not believe this story. "How could glaciers ever stretch down to the Swiss plain?" he said. "It is too warm there for ice to last."

[1]Say: shahr-pahn-TYAY.

But Agassiz decided to go over the ground and examine the evidence himself. During his next summer vacation, he set out with several friends to explore the Rhône Valley. And there he was to stumble upon one of the strangest discoveries in the world.

Exploring a Glacier

Agassiz and his friends climbed the great stream of ice known as the Rhône Glacier. Running down the length of its icy surface were dark stripes of gravel, clay, and boulders. Heaps of such loose material are called *moraines.*

The same sort of material was piled on the ground at the front of the glacier. Clearly, it had been dropped there as the ice melted.

The party hiked down the valley. Many miles from the glacier, they climbed over one heap of clay and boulders after another. Some were big hills; others were small. But all were made of the same stuff as the moraines carried on top of the glacier.

Far below the glacier front, "homeless" boulders lay scattered around. Where bedrock showed, it was polished so smooth that it glistened like silk. In some places there were long grooves in the rock. Such marks are not made by running water. The polishing and grooving must have been done by a stream of ice.

At the end of the summer, Agassiz said, "Charpentier is right. The glaciers once were tremendous rivers of ice. They filled the valleys to their brims, poured down to the foot of the mountains, spread over the plain."

An Ancient Disaster

Back at the university, Agassiz taught every day, but in the evening his thoughts often turned to the ancient ice floods.

"Just how far did the glaciers go?" he wondered. "When ice pushed below the foothills of the Alps and over the Swiss plain, the climate must have been very cold. Ice must have flowed from other mountains and buried other plains."

Agassiz pictured glaciers advancing everywhere. In his imagination he saw mammoths and other animals fleeing before a flood of ice. "Perhaps," he thought, "the mammoths and other creatures were buried alive."

A year after his glacier trip, Agassiz spoke before a meeting of scientists. They thought he would talk about his specialty, ancient fishes. But he surprised them and began to discuss glaciers.

"When glaciers spread around the Alps," Agassiz said, "the climate of Europe must have been very cold. We know that every so often in the history of the earth, whole sets of plants and animals have died out. Very likely they were killed off each time by cold—by a climate change like the one that sent the ice of the Alps pouring down over Switzerland. No doubt it was the same great cold that buried the Siberian mammoths in ice.

"Not only the Alps were ice-covered," Agassiz went on. "A long Siberian winter settled over lands once green with forests and inhabited by great beasts like those now living in India and Africa. Death wrapped all nature in a shroud of ice, which blanketed most of the continent of Europe."

"A fantastic notion!" the scholars said.

Even Charpentier, explorer of ancient glacier trails, refused to believe that ice had ever buried Europe. Yet the strange idea fascinated everyone.

"After all," the scholars said, "no one can tell just how much ground the glaciers covered. We must find out more about them."

Every summer for the next three years, Agassiz and his friends went exploring glaciers in the Alps. Other scientists also took up the study of glaciers. They made many discoveries which in time were to help clear up the riddle of the Ice Age and explain how the mammoths and other great beasts had died.

Rivers of Ice

GLACIERS of the Alps are now visited every summer by tourists. Perhaps you will travel up the Rhône Valley someday, see the glacier in action, and learn what Agassiz learned about the behavior of ice. This glacier, or one in the Canadian Rockies or anywhere else, will give you some idea of the ice floods of the past.

As you approach the Rhône Glacier, following your guide, you hear the roar of rushing water. It comes from a torrent pouring out of a cave at the base of the ice front. The cave is shaped like an archway. It is the mouth of a tunnel that runs under the glacier. Water from the melting surface drains down through cracks in the ice, flows along under the glacier, and hollows out the tunnel.

If you can get near enough to the archway, you feel gusts of cold air puffing from it. "The glacier is breathing," the mountain people say. The air in the tunnel is colder and heavier than that outside. So it flows down the tunnel and out the opening. Warmer, lighter air from outside rises into the tunnel, thaws the walls and ceiling, and in this way expands the archway. A large glacier may have an arch 100 feet high and wide enough to hold a four-track railroad.

The guide leads your party to a place beside the glacier where you can climb up on the ice. When you reach the top, you must scramble over a ridge of gravel and boulders. This is a side morainc. You see that it stretches upward the whole length of the glacier. A ridge like it extends along the other edge.

You pick up a handful of gravel from the moraine. It is wastage of a mountain. High on the crests, water seeps into cracks in the rock. Freezing there, it expands and pries away splinters and chunks, which tumble down onto the glacier.

Paths of Danger

Where the valley slope is regular, the glacier flows evenly, and its surface is smooth. But where the channel suddenly dips, beware! As the ice flows over the drop, its top layers split open. Melting widens the cracks into great crevasses.[1] Between them the ice is left standing in shapes like giant walls and chimneys. These topple and become separate blocks, some as big as houses, which cascade slowly downward.

When your party reaches a section broken by crevasses and icefalls, everyone must be roped on a line and go in single file. Then if someone slips over the edge of a crevasse, the rope will hold him and prevent a terrible fall.

In winter the crevasses fill with hard, drifted snow, which makes them easy to cross. But by summer most of the snow is melted. Only here and there enough remains to form bridges.

Inside a Glacier

You look down from a snow bridge and you can hardly believe your eyes. The gulf beneath you seems to be a piece of blue sky that has fallen into the glacier. The blue is made by light bouncing back and forth between the ice walls.

As the sun beats down on the glacier, meltwater gathers in pools. Rivulets sparkle as they flow through channels of emerald-green ice. Water tinkles musically,

[1]Say: kreh-VASS-ez.

spilling from pool to pool. Here and there a streamlet tumbles into a crevasse and flashes like a cascade of pearls. Elsewhere a torrent falls, roaring, into a funnel-like chute that it has dug down into the glacier. The chute, or *mill*, is deep. You become dizzy as you watch the foaming water plunge down, down, down. The glacier seems to have no bottom.

Your guide may mention that his countryman, Louis Agassiz, once dammed a torrent and turned it aside from its mill, then went down to explore the mill. He descended 120 feet, but did not reach the bottom. Later, other men went down twice as far into mills and found that they lead to a system of tunnels, which sometimes open into a hidden lake.

Lost Beneath the Ice

Perhaps your guide will tell you the story of Bohrer, a mountaineer who had a strange accident. One day, when Bohrer was exploring a mill, his rope gave way. Down he plunged into the heart of the glacier. The fall knocked him unconscious. When he came to, he was lying in darkness on a bed of gravel and mud. One arm pained him terribly—it was broken. He struggled to his feet and took a few slow steps, holding out his good arm to feel for some way of escape. In the murk he could barely make out surrounding walls of ice. He was in a long passage—a branch of the glacier tunnel.

Bohrer said to himself, "I'd better not go down valley. Too much water there."

So he began to pick his way upward, stumbling over boulders and slogging through mud and icy water.

After three hours of torment, he found the tunnel becoming brighter. Finally he came to a small crevasse, dragged himself through it, and got safely to the top of the glacier.

Another famous accident showed the rate at which a glacier may move. In the year 1820, a party of mountain climbers were trying to reach the summit of Mount Blanc. They chose a route up the Bossons Glacier where the going seemed easier and safer than along the rocky slopes. But climbing the glacier was dangerous, too.

One of the risks, when you get high in the mountains, is that a mass of snow may break loose and avalanche down on you. Wind, or even a noise, is sometimes enough to start an avalanche.

This is what happened to the climbers on the Bossons. A heavy mass of snow broke loose from the slope, hurtled down the glacier, and carried the climbers with it. Some of them escaped death, but three were swept into a crevasse. The avalanche filled the crevasse and buried them so deep that their bodies could not be found.

A scientist later studied the movement of the Bossons Glacier. He predicted that the bodies of the lost climbers would be delivered at the glacier front about 40 years after the accident. Sure enough, in 1861, just 41 years later, evidence of the three climbers appeared at the surface of the melting ice.

In 41 years, the remains had been carried down a height of 9,250 feet. This shows that the ice descended at an average rate of 225 feet a year, or 7½ inches a day.

How a Glacier Begins

What forces create a glacier and launch it on its march?

High in the mountains, snow drifts into a basin or hollow lying between the peaks. If the basin is above the "snow line," temperatures are so cool even in summer, that some of the drifted snow doesn't melt. More and more of it piles up, year after year. New layers weigh down on the old and pack the deep snow into grains like hailstones. Deeper still, pressure jams the grains together and forges them into crystals of pure ice. In this way, a wide, deep ice field is built up.

The mass of ice is solid, but under pressure its crystals will shift around. This sort of thing also happens to other solids. Pitch (hardened tar) is so solid that you can crack and chip it with a hammer; but if a mass of pitch is left by itself for a while, its own weight causes it to spread out flat. Lay a coin on the surface of pitch, and in a few weeks the coin will sink out of sight.

Several years ago a plane flying over the Alps crashed on an ice field above the glaciers of Mount Rose. The plane gradually settled and disappeared from sight. Now the wreckage, locked in one of the glaciers, is moving slowly down the mountain. Some time in the future it will be delivered at the glacier front, several miles from the scene of the crash.

The weight of an ice field, like the weight of any object on earth, is due to the force of gravity pulling upon it. The downward pressure on the ice tends to squeeze its edges outward. Tongues of ice are forced over the brim of the basin, through gaps between the peaks.

Creeping Stream

When an ice tongue reaches a slope, it is shoved downward by the ice field behind it and is also dragged by its own weight. The ice is meshed against the bumps and hollows of the rock bed beneath it, so it can't just slide like a big toboggan. Instead, its crystals wrench, split, and slip against one another. By making millions of tiny shifts, the mass of ice moves internally – it flows.

Great pressure is needed to force these movements. But in layers of ice right under the glacier surface, pressure is slight. This makes the upper ice less plastic, so it breaks if it is pulled or bent too much.

Along the glacier edges, where the ice is thin, it catches against the sides of the channel. The center ice keeps on pulling downslope. It splits away from the side ice, making big cracks. These widen into crevasses.

Where the whole glacier flows over a drop, the top layers crack open because they do not bend enough. Melting widens the cracks into crevasses. The deepest of these do not go down much further than 200 feet. Below that depth, the great pressure holds the ice together, no matter how much the ice may twist and bend.

As a glacier flows along, it plucks loose chunks of rock from its bed. They lock into the ice like the teeth of a carpenter's rasp. Driven by the moving ice, the teeth gouge the bedrock and wear it down. The scrapings are picked up by the ice and carried along with all the other rock fragments. The whole load moves down toward the glacier front.

New ice, always flowing from the heights, supplies the glacier and shoves it along. The giant moves so mightily, it seems no force can stop it from plowing on and on.

But the giant has an enemy as strong as itself — the warmth of the lowland. As the glacier pushes down valley, its front or "snout" begins to melt. It shrinks faster and faster. Even though new ice is always streaming down, the snout is brought to a halt by melting. If the climate is steady and the ice supply regular, the snout stays in one position.

At times in the past when the flow of ice overbalanced melting, glaciers advanced. Two hundred years ago, they were gaining ground everywhere. In Norway and Iceland, glaciers overran many farms. People saw their pastures buried and their houses crushed and carried away.

During the past hundred years, the world's glaciers have been shrinking. As those in southwest Greenland retreated, they uncovered the foundations of ancient Norsemen's homes.

In Alaska, some glaciers that once flowed into the sea now end many miles inland. In the Rocky Mountains and the Sierra Nevada, many glaciers have disappeared. Others have retreated so far that only stumps remain high in the mountains.

Tracking a Giant

BELIEVING that glaciers had once covered Europe, Louis Agassiz went to northern Germany and Poland to look for their traces. He found moraines and glacial boulders scattered over many hundreds of miles.

Agassiz saw that his guess about the glaciers had been right. An enormous ice sheet must have covered the northern half of the continent. As the ice flowed southward, it brought along a load of rock fragments. Later, while retreating, it left this material behind. In places where the retreating ice front paused for a while, the boulders and dirt piled up to form moraines. But where the ice retreated steadily, the material was spread out in a layer.

Foreign rocks are found almost everywhere in northern Germany and Poland. A few are as big as railroad cars. They match the rock of Scandinavian mountains which lie several hundred miles north of of the Baltic Sea. The "foundlings" show that ice streamed right down to the area now covered by the Baltic, crossed over it, and plowed into central Europe.

To North America

"What of North America?" Agassiz wondered. "If Europe was cold enough to be buried under ice, could the climate have been very different anywhere else in the northern half of the world?"

Agassiz sailed for America, and his ship landed at Halifax, Nova Scotia, in the autumn of 1846. Later, telling about his excitement, he said, "I sprang on shore and started at a brisk pace for the heights above the landing. There I found the familiar signs, the polished and grooved rock faces so well known in Europe."

From his first glimpse of North America, Agassiz was sure that this continent, too, had been overrun by ice sheets.

He became a professor at Harvard University, lived in Boston, and from there made trips around the New England States. He went to the White Mountains in New Hampshire and the Green Mountains in Vermont, where he discovered hosts of moraines, polished rock faces, and foundlings.

Following the Trail

Agassiz traced moraine hills along the road from Springfield to Boston. He saw that the islands in

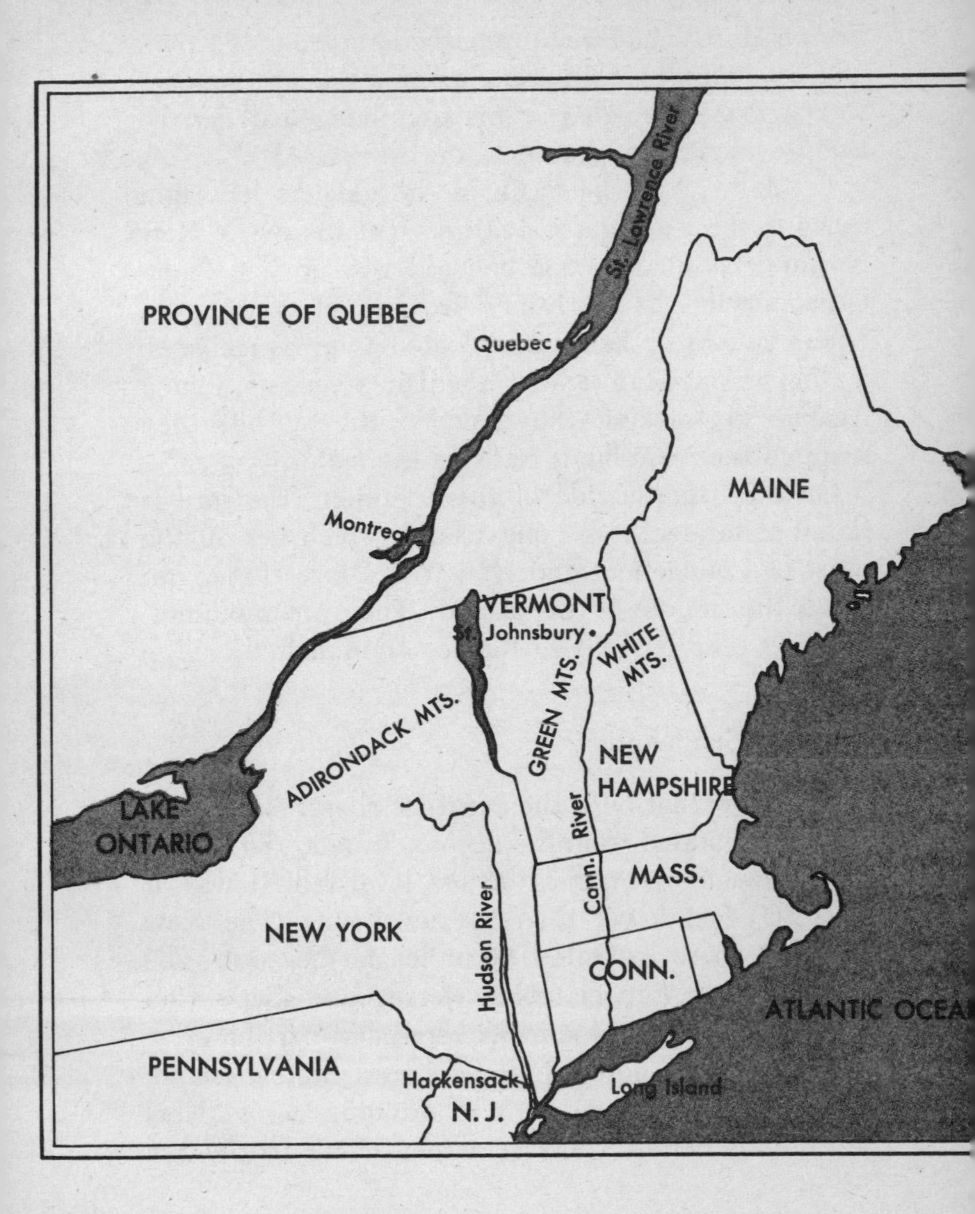

St. Lawrence River
PROVINCE OF QUEBEC
Quebec
MAINE
Montreal
VERMONT
St. Johnsbury
WHITE MTS.
GREEN MTS.
ADIRONDACK MTS.
NEW HAMPSHIRE
LAKE ONTARIO
River
Conn.
MASS.
NEW YORK
Hudson River
CONN.
ATLANTIC OCEA
PENNSYLVANIA
Hackensack
Long Island
N. J.

Boston Harbor had been smoothed and rounded by a glacier. Since he found no moraines near the city, Agassiz concluded that in this area the ice had carried its load of dirt and boulders out to sea.

In Maine, too, the track of the glaciers led right down to the sea. Agassiz realized that the whole New England coast had been covered with an ice sheet. Great glaciers had streamed from it into the ocean, where chunks broke off and floated away as icebergs.

Many American and Canadian scientists joined Agassiz in tracking the glaciers. Bit by bit, they mapped the wide-flung trails of the ice.

On the Atlantic side of the continent, the trail is found as far south as Long Island, which lies off the coast of Connecticut and New York. Two ridges run down the middle of the island. They are moraines, marking positions where the ice front halted.

Ice Age Geography

The snow that built the great ice sheets came from water evaporated from the oceans. In fact, the oceans lost so much water that the sea level fell. It was at least 200 feet lower than the level now. The New England shore extended 80 miles farther out. The ocean bottom between New Jersey and Cape Cod was dry land, part of it ice-covered and part bare.

The southernmost moraine crosses Staten Island, and you find sections of it continuing across New Jersey and Pennsylvania. You can climb to the top

The American mastodon of the Ice Age

of the moraine, pick up a handful of stony dirt, and say, "The ice brought this here."

When the ice sheet buried the site of New York, the land to the south and west was free, and herds of reindeer roamed over it. There were also musk oxen, mammoths, and their relatives the mastadons. Giant ice rafts floated along the coast as far south as the Carolinas. Seals and walrus swam offshore. Sometimes they climbed up on the ice rafts and lay basking in the sun.

Today, chains of moraine hills stretch westward all the way across the continent. They show how vast the ice sheets were. Imagine glacier snouts joined for 3,000 miles. Picture greenish-blue cliffs washed by lakes of meltwater. Hear the roar of torrents pouring from archways along the base of the cliffs. Think of an ice mass so heavy that the solid crust of the earth bent down beneath it.

When the ice retreated northward from New York, the earth's crust in this area rose again. This made the land slope upward away from the ice, forming a basin along the ice front. Rivers emptied into the basin, and the moraines looping around its outer edge helped to hold in the water. In this way a big lake was impounded in the place where Long Island Sound is today.

As the ice front melted farther back toward the north, the lake bed rose. Rivers carried in sand and clay, which settled over the bottom. The lake became shallower and shallower. The remaining water drained out through breaches in the moraines. Where the lake had been, there was now a dry bed of clay, crossed here and there by streams. Meanwhile new lakes formed in the north, along the retreating ice front.

Retreat of the Ice

The story of the retreat was pieced together gradually by a number of scientists. One of them, Ernst Antevs, came from Sweden to investigate the area of New York and New England. He scouted around in the neighborhood of the southernmost moraine. At

Hackensack, New Jersey, he found what he was looking for — clay pits with high walls. The material in these walls had once been sediment at the bottom of a lake.

Material deposited in glacial lakes is not jumbled like material dumped by ice. It is sorted according to size. You can easily see how this happens. In a glacial river the current is strong enough to carry sand, along with finer particles of silt and clay. Where the river enters a lake, the current slackens and the sand drops to the bottom right there. Silt spreads through the lake; then it, too, settles. But tiny bits of clay continue to float in the water for months.

During winter, the river shrinks. Flowing very feebly, it carries hardly any new material. Meanwhile the bits of clay floating in the lake gradually settle

as a fine covering over the coarser summer deposit.

This must have happened in Lake Hackensack, for the material of the old lake bottom was arranged in neat layers. Each had a coarser part — the summer deposit; and on top of this a finer part — the winter deposit. Each layer marked one year.

History in Layers

It would have been nice if Antevs had found 50,000 layers one on top of another, recording 50,000 years of the earth's history. But the lakes did not last that long. So Antevs looked for a series of clay beds that had been laid down in one lake after another.

He chose the Connecticut Valley as his route and started by counting layers in beds around Hartford. From there he worked toward the north, going as far as St. Johnsbury, Vermont — a distance of 185 miles.

When Antevs had picked out a clay bank, he would trim the face with a spade and trowel. Then he fastened a long strip of paper on the bank from top to bottom and marked the thickness of each layer on the paper.

A very thick layer meant the summer of that year had been warm. A lot of ice melted, the rivers were swift, and they washed a great deal of silt and clay into the lake. A thin one meant the summer had been cool. Less ice melted, the rivers were smaller and slower, and they carried less silt and clay.

Because of these differences in the thickness of layers, Antevs could compare layers of one locality with those of a neighboring place. In each deposit he would find a matching series of thin and thick layers. They belonged to the same years, so Antevs did not count them twice. He began counting a new set of layers where an older one left off.

Antevs counted 4,000 clay layers between Hartford and St. Johnsbury. This showed it took 4,000 years for the ice front to retreat the 185 miles between the two places. The average rate of retreat was a mile in 22 years, or 238 feet a year.

At this rate the whole retreat, from the southernmost moraine to northern Canada, would have taken about 35,000 years.

Long as this period seems, it was only the last part of the history of the ice sheet.

Back to the Beginning

The beginning of the story is harder to trace, but if you were an expert you could find out several important things about it. Suppose you were on the Long Island moraine and picked up some stones made of Adirondack Mountain rock. They would show you that the ice which carried them came from the Adirondacks, in northern New York State.

If you went to the Adirondacks, or to the White or the Green Mountains, you would find foreign boulders there too. Some of them match rock in the Province of Quebec 500 miles to the north. This is remarkable, for during the Ice Age, as today, the wide, deep valley of the St. Lawrence separated Quebec from

New England. The boulders show that ice from central Quebec crossed the valley. The ice must have poured into the valley for centuries, until it was filled. Then new layers moved over those beneath.

In order to reach the mountains to the south, the ice had to push uphill from the Quebec lowland. Layer piled on layer, each new one shoving higher up on the slopes. Ice tongues flowed on southward through gaps between the ranges and plowed down the Connecticut and Hudson valleys. When the flood grew high enough, it streamed over the very summits of the mountains.

The Quebec ice sheet must have resembled the ice sheet that covers Greenland today. It had the shape of a broad dome, high in the center and low around the edges.

It began to form when the climate of central Quebec turned cold, and snow and ice piled up. Gradually, a sprawling mountain of ice arose. Its weight and pressure squeezed the edges out in all directions.

A second ice dome formed in the Keewatin region west of Hudson Bay. In time, the fronts of the Keewatin and the Quebec sheets met and then spread as one.

In western Canada, a third ice dome piled up on the plateau between the Rocky Mountains and the coast ranges. During the height of the Ice Age, the fronts of the Western sheet and the Keewatin sheet met at the base of the Rockies. But for long periods the fronts lay some distance apart, and a long corridor from north to south was left open between them.

Ice of the Quebec and the Keewatin sheets pushed down the Mississippi Valley as far as the place where St. Louis stands today. Scientists have tried to figure out how thick the ice domes had to be in order to shove their fronts that far. They think the Keewatin dome was more than three miles thick.

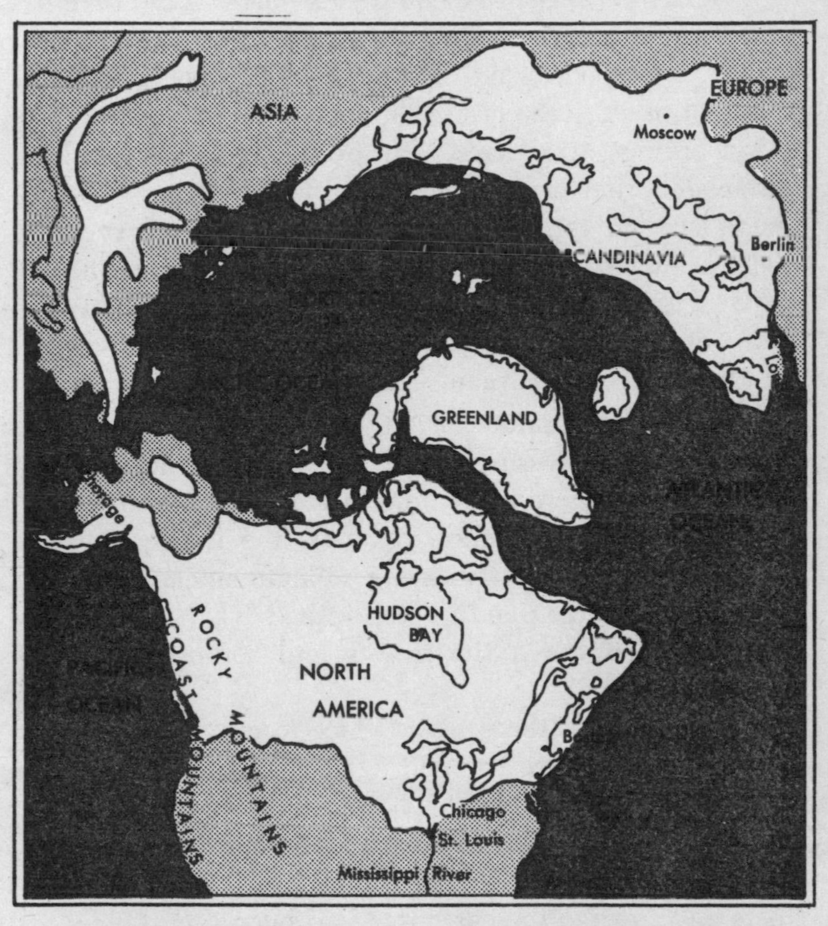

The white areas show the greatest extent of ice sheets in the northern parts of Asia, Europe, and North America.

At the time these ice sheets were spreading in North America, Europe was overrun by an ice sheet streaming from the mountains of Scandinavia. Together, the ice of the two continents buried more than seven million square miles of land.

The ice floods have spread and retreated four different times in the last million years. Scientists have made borings at a place near Berlin, Germany, and discovered that several beds of glacial material lie there, one on top of another. Sandwiched between them are layers of peat, which is made of plants and parts of trees pressed into a solid mass. This shows that between the cold periods came times when the ice disappeared and plants grew over the land.

Each glacial period lasted tens of thousands of years, and the warm periods between them were longer. There were really four ice ages, though we speak of the whole million years as "the Ice Age."

After a period of warmth, the climate again turned cold. Domes of ice piled up and spread. Their moving cliffs battered down the forests and scraped away every green thing.

Animals fled, if there was a place to go. Sometimes bands of them were trapped against high mountains or the sea. The pockets where they lived became smaller and smaller. There was not enough food, and the animals starved. In time the moving ice buried them and covered the earth that had been their home. Nothing remained but a lifeless desert of ice and snow.

The Tundra and Its Creatures

BECAUSE an ice sheet is white, it helps to make a cold climate colder. A white surface doesn't soak up much heat. That is why we wear light-colored clothing in summer. White clothes bounce the sun's rays away from us.

In the same way, a cover of ice and snow bounces the sun's heat away from the earth. The ice stays cold and chills the air above it.

When air cools, it shrinks. Then there is more air in the same space, so it is heavier. In the weatherman's language, its pressure is high. Over an ice sheet, the high pressure of the air causes it to shove down and outward, forming winds that blow from the ice.

These winds are very cold and therefore dry, for cold air holds little moisture. The winds make a climate in which plants that love warmth and moisture cannot grow. When ice sheets covered northern Europe and North America, the southern parts of the continents became too dry for trees. Forests died out and were replaced by grasses, which get along with less moisture. Nearer the ice fronts, a belt of cold-loving plants stretched across Europe, Asia, and North America.

What Is a Tundra?

You can get an idea of this belt from the similar one that encircles the Arctic today. Toward the northern rims of the continents, the forest dwindles to patches of dwarf trees. Then even the scrub comes to an end, and a treeless plain stretches out, flat and bleak, to the horizon. This is the tundra. The only seasons here are winter and summer – a winter of ten months and a summer of two. The cold, dry winds bring little snow and rain. In a warmer zone, the land would be desert.

But here it is so cold that moisture is stored in the forms of ice and snow. The very ground is frozen. In summer the top layer thaws, but beneath it the earth remains ice-locked, as it has been for thousands of years. The frozen soil is as hard as rock. It bars roots from growing deep, so trees cannot live here.

Water does not drain through the frozen ground, but lies over it, forming many lakes, ponds, and bogs. Waterlogging of the soil makes it sour. Only certain kinds of plants can live in sour soil, but they grow abundantly. Hardiest are the lichens – little, slow-growing plants that can even cling to rocks and stay alive. Mosses are plentiful, and there are sedges, bog berries, heather, and some grasses. A relative of the willow, not more than three inches tall, creeps along the ground.

Animals of the Tundra

The tundra is the home of the musk ox and reindeer, which live by feeding on the lichens and other plants. The musk ox has been wiped out in the Old World, and few remain in Greenland and North America. No wonder this animal has nearly become extinct — it is too brave for its own good. When attacked, the herd does not run. Instead, the six or eight adults form a circle around the young and face the enemy. This defense works well against wolves and other animals. If a wolf dares to run in close, the musk oxen gore it to death with their horns.

But the defense circle is useless against hunters armed with guns or with bows and arrows. Hunters

used to set a few dogs on a herd to hold it at bay, while they stood at a safe distance and shot as many animals as they wished. Now musk oxen are protected by law.

Reindeer, which run from their enemies, have survived in large numbers. In arctic Asia and Europe, people raise tame deer for their meat, milk, and skins. But in North America the native reindeer, or caribou, as they are called, are wild.

The caribou are great travelers. In winter, most of them wander south to the edge of the forest. There they find shelter from the cruel winds, and they can dig up plants from under the snow.

When winter ends, the caribou return from the forest to the tundra. Some of them travel as far as the straits separating the continent from the arctic islands. The deer seem to know just where they want to go. Fearless, they venture on the ice that still covers the straits and gulfs. They journey on and on, until they have left the land out of sight behind them. Somehow they find their way to their summer home. This is Victoria Island, where lichens are plentiful and cool winds keep away the flies.

When ice sheets covered the present homes of the reindeer and musk oxen, the zone of tundra lay farther south. It crossed what is now the middle of the United States. In Europe, it crossed France, Germany, Poland, and Russia. Reindeer, musk oxen, and woolly mammoths roamed this zone on both continents. Their companion, the woolly rhinoceros, ranged over Europe and northern Asia, but never came to North America. With its massive body and long, slender horn, the great beast must have looked like a modern African rhino, except for its fur.

The tundra home of these animals shifted every time the ice fronts moved. When the ice advanced to the south, its cold, dry winds went before and made a climate where tundra plants could flourish. Lichens, mosses, and heather replaced grasses and trees.

When the ice retreated, the tundra plants crept back toward the north, covering the new moraines. Wherever the tundra shifted, its herds of reindeer, musk oxen, mammoths, and other creatures shifted also, in order to remain in the kind of homeland that gave them their food.

Animal Wanderers

DURING the Ice Age, Europe, Asia, and North America had more large animals than Africa has today. Several kinds of mammoths, with their cousins the mastodons, roamed the lands of the northern hemisphere. The imperial mammoth, which ranged over the southern part of North America, stood 17 feet tall. If it existed today, this great beast could easily look over the head of an African elephant.

When the sea level was low, animals could travel freely between North America and Asia. The continents were joined by land where Bering Strait separates them today. Even now, the waters of the Strait are very shallow. If the sea bottom were to rise only 120 feet, a passage of dry land would connect the continents.

The route was usually free of ice. In northeast Asia, winters were so cold that the air held little moisture. Snowfall therefore was slight and formed only small glaciers.

In Alaska the winters were somewhat milder, but the high mountains along the coast caught most of the moisture brought by winds from the Pacific. The snow fell mainly on the slopes facing the ocean, and

ice forming there flowed harmlessly into the great sea.

The dry air passing over the mountains brought little snow, so ice did not spread over Alaska. The Yukon Valley was free of ice and opened into the corridor between the Keewatin ice sheet and the Rockies. Thus a long, ice-free route led all the way from Asia to the plains of the American Southwest.

Great Migrations

Animals migrated both ways from continent to continent. The bison, mammoth, and mastodon families came from Asia and spread to North America. The camel, a native of North America, traveled to Asia.

A wonderful assemblage of animals lived on the North American plains. Horses of several types roamed in enormous herds. Some were ponies; others grew

larger than any horses of today. With their trim, long-legged bodies, built for speed, they could travel far in search of grass and could outrun enemies. When a herd was frightened and stampeded, the plain must have echoed with the thunder of their hoofs.

Wherever plant-eaters go, flesh-eaters pursue them. Wolves, lionlike cats, bears, and other beasts of prey migrated with the plant-eaters from continent to continent.

The Terrible Sabertooth

Among the flesh-eaters was the great cat called the sabertooth, because of the tusks that jutted from its upper jaw. They were not round, but were flattened like blades and curved backward. The whole body of the creature was built in a way to put power behind those terrible weapons. Its neck, shoulders, and front legs were thick and heavily muscled. When the beast leaped upon its prey, it did not bite. It stabbed the victim with its deadly "sabers."

The last branch of the family in America has been named the California sabertooth because its remains are plentiful there. Many of the beasts were trapped and buried alive in tar pools at a place that is now inside Los Angeles.

Trapped in the Tar Pools

Oil seeped up from rocks under the tar pools, drenching the earth and sand. The oil contained asphalt, which made the ground sticky. Sometimes water lay on top of it. Unsuspecting animals came there to drink, and when they stepped on the gummy bottom, it caught and held them. They struggled to pull themselves out, but only sank deeper.

Bones of the trapped animals were preserved by the asphalt. Many skeletons have been taken from the tar and now stand in our museums. Among them are several types of horses, long-horned bison, camels, and that strange beast, the ground sloth.

Ground sloths came from South America. The last type to arrive was as bulky as an elephant. It had a thick, powerful tail. It used the tail as a prop when it squatted on its haunches, stripping leaves from trees. Flesh-eaters probably dared not come within reach of the giant's forepaws, for they were armed with great claws.

When a sloth or other animal was caught in a tar pool, its struggles attracted flesh-eaters, but what seemed an easy meal turned out to be a baited trap. When wolves and sabertooths came to attack the victims, they themselves often sank into the mire.

Giant vultures, the largest flying birds that ever lived, hovered over the tar pools, waiting for trapped animals to die, then settled on the carcasses for a feast. Some of them ate so much that they could not fly away, and they too sank into the sticky earth.

Most kinds of large animals spread all the way across the continent. On the east coast, animals migrating southward reached Florida and were stopped by the sea. Many settled, and their descendants stayed during the ice periods and the warm periods too.

Tapirs, whose modern cousins inhabit the jungles of South America, splashed along the watercourses, feeding on plants that hung over the banks. Ground sloths half sat and half stood, grasping tree branches and eating leaves endlessly, as they had to do in order to nourish their enormous bodies.

Among the plant-eaters were horses, asses, camels, and deer. Wolves hunted them, and so did jaguars and sabertooths. But all these animals made way for the lordly mastodons and mammoths when they came crashing through the thickets gathering trunkfuls of leaves.

The Great Hunters

SKELETONS of Ice Age animals are plentiful, but skeletons of Ice Age people are rare. This is explained simply. Animals blundered into quicksand or tumbled over riverbanks. Their bones were buried under mud that preserved them. But people were too intelligent to fall into such traps.

Human remains would be more plentiful if people of the Ice Age had buried their dead. Most of them, it seems, left their dead in the open, where the bodies were destroyed by animals and by decay.

The Hand-Ax People

Human beings, however, left traces more durable than bones. They used tools, some of which were made of flint and other hard stones. Many of the tools have been dug up in Europe. The simplest type is an oval stone chipped on one side so as to give it an edge. The other side is left round, and will fit into the hand. Such a tool is called the hand ax. A person holding one can use it either as a weapon or as a cutting and scraping tool.

"Hand-ax people" originally lived in Asia and Africa. About 300,000 years ago a group of them drifted into Europe. A few of their bones have been found — just enough pieces of skulls and jaws to tell us these human beings were low-browed, big-jawed creatures.

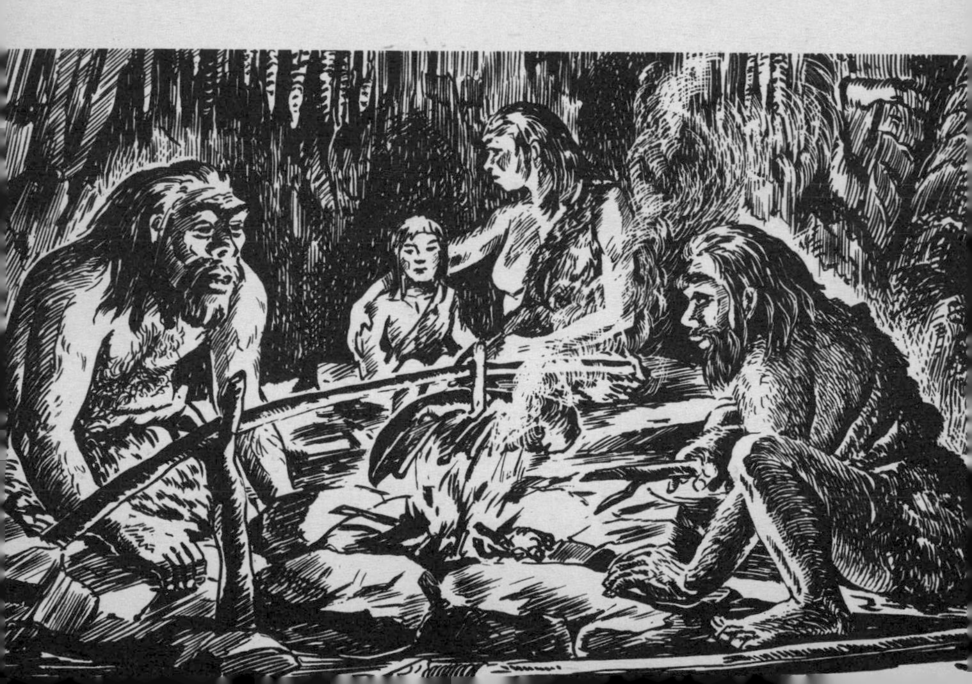

The hand axes are usually found in the banks of streams. It seems the people roamed along the rivers, living on clams, mussels, and roots, which they dug up with their stone tools.

During the time of the hand-ax people, the second ice sheet disappeared from Europe, and forests crept northward. In the forests lived African animals like the hippopotamus, elephant, and rhinoceros.

When the climate turned cold again and a new ice sheet spread, the hand-ax people were driven away. Very few could find places with enough shellfish and roots to keep them alive.

Neanderthal Men

While the third ice sheet was retreating, people of a new type appeared in Europe. At times they took shelter in caves, and some of them buried their dead. In several places bones have been found preserved in graves and in cave dwellings. Some were dug up in the Neanderthal district of Germany, so the people have been named "Neanderthal men."

A Neanderthal man was short-legged, compared with us, but not small. He had a barrel chest, powerful arms, a thick neck, and a large head and face. He slouched forward when he walked. A great ridge over his eyes must have made him resemble an ape.

The Neanderthal men used hand axes that were expertly chipped for different uses like cutting, sawing, and scraping. They certainly had spears and other weapons, for they were hunters of large animals. We know this because bones of bison, horses, and reindeer

Tools of Neanderthal man

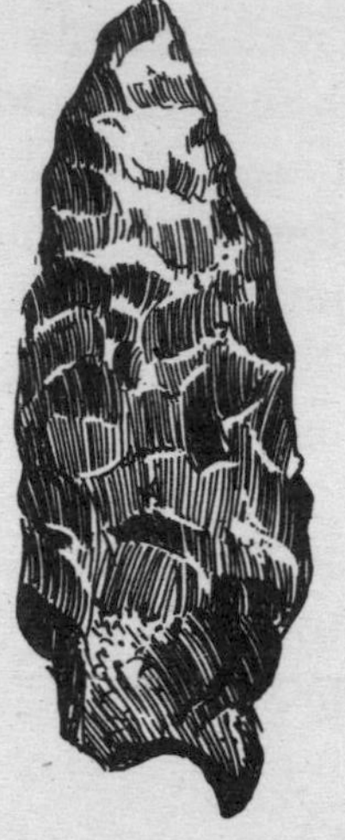

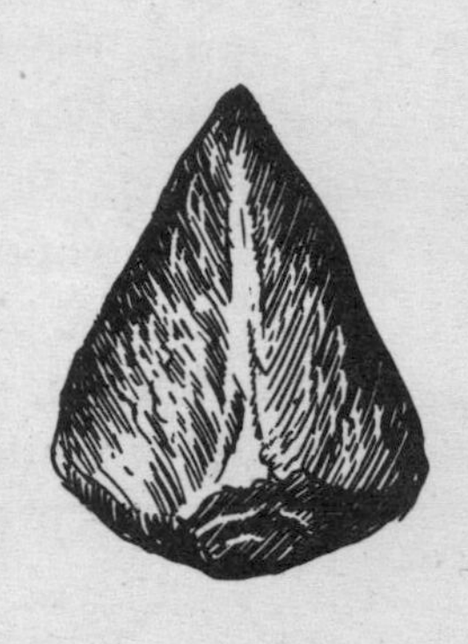

have been found with Neanderthal remains.

Lions, bears, and wolves were the people's rivals. They chased the same game. Some of them took refuge in the same caves. There was a cave bear, a cave hyena, and a cave lion. Sometimes the people came face to face with these beasts in the caves and fought them with blazing torches, clubs, and spears.

When the fourth ice sheet – the last – spread over Europe, the Neanderthal people disappeared, probably because they could not make a living.

The Coming of the Great Hunters

Later, while the fourth ice sheet was retreating in northern Europe, tribes of mighty hunters appeared in the south. We think they came there from Asia and

Africa, following the big game animals. The newcomers were like us in build. Judging from the skulls which have been found, some resembled modern Europeans, some Africans, and others Eskimos.

Several branches of the human family kept alive during the centuries of trial. Through their struggle for life in the changing world, human beings themselves changed. The groups that became most skillful in hunting and working together prospered and spread to new places. The future belonged to them.

Tools and weapons left by the Great Hunters show that they were very clever people. They fashioned many kinds of implements from wood, flint, and ivory. Some genius invented a wonderful weapon, the harpoon. This was a spear with a detachable point that carried a line. When the point hooked into an animal, the Hunters could hold it by the line.

The Great Hunters spent their summers following game on the plains and tundra. In winter they sometimes took refuge in caves, living in the entrances, which were light and airy.

Bone needles and needle cases, dug up from campsites, show that the people sewed material for garments. No doubt they learned to clothe themselves with animals skins, as the Eskimos and other arctic peoples do today.

Since no trees grew on the plains and tundra, the Hunters had no wood for fires. But they discovered another kind of fuel. We know about it from stone lamps, shaped like trays, which have been dug up. Several of these lamps had bits of animal matter clinging to them. When tested, this stuff proved to be remains of fat. A wick of some material like moss was set into the fat and lighted.

You can imagine a group of fur-clad men, women, and children gathered in their cave shelter, protected from winds that howl outside. They squat in a circle around a lamp, their faces lighted and warmed by the bright flame. They feast on chunks of tasty meat, and other chunks hang over the lamp, roasting. Life is good, in spite of the cold and ice.

Secret of the Caves

HAVE you ever tried to draw a horse? It isn't easy, is it? The Great Hunters painted beautiful pictures of horses and other animals. You can see them, still fresh and lifelike, on the walls of caves in southern France and northern Spain. The figures are so wonderfully drawn that you might think the artists had tame animals to pose as models. Reindeer are shown grazing, running, or swimming across streams. Wild oxen are fighting. Bison fall, wounded by the spears and harpoons of hunters.

The pictures tell us that wild horses, asses, bison, and antelopes roamed the grassland of southern Europe. We know the tundra was not far away, for the artists painted woolly mammoths and woolly rhinos, and their favorite subject was the reindeer. Bones of tundra animals and plains animals have been dug up from the same campsites.

During winter, herds of reindeer, mammoths, and musk oxen migrated to the plains of southern Europe. In summer, they returned to the tundra. Bands of the Great Hunters followed the reindeer northward, all the way to the ice front.

The Chase

We can imagine the wild country the Hunters traveled over. First they crossed old moraines covered with plants, then new ones, bare of any green thing. Rivers wound between heaps of boulder-strewn earth. Where the streams were dammed by moraines, wide lakes spread.

As the Hunters closed behind the fleeing deer, they saw the ice front ahead of them. Torrents roared from caverns at its base, filling a lake. Ice cliffs rose from the water and thrust their gleaming blue towers against the sky.

The Hunters drove the deer to water in order to slow and tire them. Then they could be harpooned. We are able to guess this from one of the cave paintings in which deer are shown swimming with only their heads above water. The artist must have learned to paint swimming deer when he hunted with his band and chased the animals to water.

You can be sure that every able-bodied person, young and old, was needed for a big game drive. Women and children carried torches, waving them to frighten the animals and chase them toward the harpooners and spearmen.

Sometimes the Hunters drove their game over cliffs. At a place called Solutré, in France, there is a steep cliff at the foot of which thousands of skeletons of horses have been found. When surrounded, the terrified animals plunged over the edge of the cliff and were killed. Then the Hunters had only to gather the meat at the foot of the cliff.

The woolly mammoth, king of the tundra, was too mighty to be chased. When a hunting band came upon a herd of mammoths — or when a herd came upon them — the people quickly got out of the way.

But the Hunters found a way to trap the king of beasts. They would dig a pit along the trail, cover it carefully, and wait for a mammoth to fall into it. Then they surrounded the beast and attacked it with spears.

Place of Mystery

If you study the paintings of the Great Hunters, you can learn the secret of their skill and courage. The paintings are located, not in the cave entrances where the Hunters camped, but in deep, dark, distant passages. Why, you wonder, did the artists hide their work?

The practices of other hunters suggest an answer. In recent times, Australian tribesmen had caves where they kept figures of game animals. They thought that the living animals were connected with their pictures. A part of them, their "soul" or "life," was supposed to be in the pictures. By guarding the pictures, the tribesmen sought to protect the animals and control them.

The life substance of the people, they said, was stored in the same cave. The tribesfolk believed that animal souls were the same as human souls, and called the animals their brothers and sisters. The tribesmen held ceremonies in the cave to keep the animals plentiful and the hunting band strong.

Knowing of such customs, you can try to guess the

meaning of the Ice Age paintings. You can imagine a ceremony that the Hunters may have held in the depths of their caves.

Very likely, the ceremony was mainly for the young people. The boys and girls no doubt had heard of a mysterious, secret place, but dared not speak of it aloud. They whispered about the time when they would be taken there for a strange ceremony. They awaited this event eagerly, but with some fear.

Cave Ceremony

The great day came. A man or woman carrying a torch led the boys and girls into the mysterious cave. The leader crouched and squeezed into a low passage. The young people followed, one by one. They found themselves in darkness, crawling through a narrow gallery.

Drums began to boom. The gallery widened out. As each youth stepped from it, he was blinded by a flood of torchlight. He found himself in a great chamber, and what he saw there made him tremble.

Bison, about to charge, glared with fiery eyes. Horses, oxen, deer, and rhinos crowded near. Above the other animals towered the mammoth, ruler of the tundra. The beasts seemed real, but they were paintings. The flickering torchlight made them appear to move, as if alive.

Men and women crouched in a circle. Swaying their shoulders, they chanted to the music of the drums.

They sang of their great secret, in words perhaps like these:

> Behold our kin of tundra, plain –
> The horse, the deer, the mighty mammoth.
> We keep their life in safety here;
> For each we kill, another comes.
> We hold them to the hunting ground.
> We guard their life, their life is ours.
> Our strength is great, for we are one,
> Kindred of the hunting band.

By joining in the ceremony, the boys and girls became grown-up members of the band. They learned that all during their lives they would work with the whole group for the good of all.

This was the secret of the caves, the secret of the Hunters' strength. They helped one another in game drives and in all the tasks of their life. The dangers they faced were great, but the Hunters saved themselves. Because they knew how to work together, they lived through the Age of Ice.

Pioneers of the New World

FOR thousand of years the summers were warm. The ice sheets retreated in Europe and North America, uncovering land farther and farther to the north. Around the melting ice fronts lay a wilderness of bare moraines, lakes, and dry lake beds. Gradually, the moraines and clay beds turned green with plants spreading over them. As the zone of tundra crept northward, so did the zone of grassland behind it, and then the forest.

Animals of each zone moved northward to stay in their own kind of home. The reindeer, the woolly mammoth, and the musk ox migrated all the way to the shore of the Arctic Ocean.

Where the game herds of Europe and Asia went, the Great Hunters followed. Their migration lasted for many centuries. The people had no written history, so later generations probably did not know that their ancestors had come from places far away. Surely none of them dreamed that they themselves were to discover the shores of a northern ocean, and some were to reach a new continent.

The First Dogs

Somewhere, in the course of their wanderings, hunters of Asia adopted a very important animal friend, the dog. We can only guess how this happened. All over the world, in later times, hunting people had dogs. Usually they were more than just pets. Sometimes they were set to trailing and chasing game. This must have been their first work, for dogs are hunters by nature. That great hunter, the wolf, is just an untamed dog.

It is hard for us to imagine people ever taming wolves. Today men fight wolves as enemies, and the wolves in turn act as enemies. They run away from the terrible human creature.

This has not always been true. Long ago, wolves gathered around hunters' camps to feed on their scraps. People tossed them bones, and because they were treated as friends, the wolves grew more and more tame. When you look at an Eskimo dog, with its sharp muzzle and thick fur, you don't doubt it is descended from the wolf.

Discovering America

About 15,000 years ago, bands of hunting people reached the northeast tip of Asia. At that time Bering Strait, which separates Asia from North America, was narrow and shallow. Islands rose here and there across it like steppingstones. In winter, ice connected the islands and made a dry route between the continents.

The hunters stood on their home shore, looked across the waste of ice, and saw, dimly outlined, a far, strange land. Reindeer sometimes crossed the ice going toward that land or coming from it. Herds no doubt roamed there just as they did at home. But did people dwell there? The hunters must have talked about the unknown country for a long while. Finally they made up their minds to explore it.

They packed their food and weapons, bundled up

the small children, and accompanied by their dogs, set out to cross the icy Strait.

When they reached the other side and stepped ashore, they saw game but no people. It was truly a new world, for until then no human being, so far as we know, had ever set foot upon North America.

The discoverers, who were ancestors of the American Indians, traveled along the Yukon or the Mackenzie Valley. Some turned southward, following the old route of migration east of the Rockies, where hosts of animals had passed. They found it was a wonderful hunting ground, inhabited by great herds of reindeer, mammoths, bison, and musk oxen. Bands of people in time wandered south as far as the plains. There they hunted camels, southern mammoths, horses, and long-horned bison.

The Find at Dead Horse Gulch

Until the 1920s, nobody knew for certain that people had lived on the plains during the close of the Ice Age. Then a discovery was made by a black cowboy who worked on a ranch near Folsom, New Mexico. He was riding through a ravine called Dead Horse Gulch, looking for stray cattle, when he noticed some white objects sticking out of the bank.

The cowboy rode up to the bank to investigate. He saw the things were bones and pulled some of them out. They seemed to be buffalo bones, but the horns were long and straight, not stubby and curved like the horns of ordinary buffalo.

This was so strange that the cowboy forgot about

the strays he had been looking for and kept on digging. He saw something—picked up an arrowhead, then several others. But were they arrowheads? The flint points were different from any arrowheads the cowboy had ever seen. They were long, thin blades, with a groove running down the middle of each side, as on a bayonet.

The cowboy told all his friends of his discovery. Scientists heard about it and came to investigate. Digging into the bank, they took out everything they could find. There were skeletons of twenty-three long-horned bison. Among the bones lay nineteen flint points and also a flint scraper and a knife. Clearly, this place had been the scene of a bison kill.

Only one thing was missing, or twenty-three things —the tails of the bison. Hunters could tell what this meant. "The tail goes with the hide," they say, meaning that a hunter, when skinning his kill, leaves the tail attached to the hide. The flint scraper and knife showed that the hunters had skinned the carcasses and cut up the meat right on the spot.

The Bison Hunters

What sort of people were these bison hunters? Did they resemble the Indians who hunted buffalo on the plains at the time when Spanish explorers first visited the Southwest? A companion of the explorer Coronado wrote a description of these Indians and their life. The chronicler says that a band of people would attach themselves to a bison herd and practically become a part of it. Men, women, and children followed the herd wherever it went, and carried all their possessions with them. Each dog, too, carried a pack or dragged a sledge made of poles.

The band depended on the herd for their food and all other necessities. Bison hides were their tent coverings and blankets. Bison sinews served them as thread for their skin clothing. Their tools were made of bison

horn and bone, their water bags of the stomachs. They used dry bison dung as fuel for their campfires. This way of life was like the existence of Ice Age people. The Plains Indians possibly were descendants of the hunters who killed the bison at Dead Horse Gulch.

Animals That Died Out

Since the discovery of the flint points at Dead Horse Gulch, others like them have been found in localities all the way from Alaska to Mexico. A point of this type, known as a "Folsom point," is skillfully chipped, giving it a keen edge. It is grooved down the middle of each side, which makes it thin and very good for piercing.

At a place called Lindenmeier, in Colorado, bones of a long-horned bison were dug up, and a broken Folsom point was found wedged between two bones of the spine. The bones were tested and found to be about 10,000 years old.

Several years ago a rancher was plowing a field in Roberts County, Texas, when his plowshare turned up some ancient bones. The spot was excavated and turned out to be an old water hole where southern mammoths came to drink. A number of mammoth teeth and bones were found, and near them lay three flint points and a scraper.

The earth at this place told a story of climate change. The layer that had been at the bottom of the water hole contained a sprinkling of black stuff — remains of plants that once grew around the pool. On top of the old pool bed was a deposit of wind-blown dust. The mammoth bones were found in this material. Clearly, the country had been drying up. Drought and man were the two enemies that killed off the southern mammoths.

Other relics have been found in caves in the Sandia Mountains of New Mexico. Flint points and tools lay among the bones of mammoths, ground sloths, horses, camels, and long-horned bison. Hunters evidently carried the carcasses into the caves for storage.

Bones of ground sloths, camels, and wild horses have been tested, and the results show that these animals still existed in the Southwest about 8,600 years ago. We don't know just when and how the last of them died out. But certainly drought and shortage of food cut down their numbers, and hunters helped to wipe out the rest. Horses disappeared along with the other animals. No horses were to roam the American plains for several thousand years, until the Spaniards came with their horses from Europe.

Mammoth Hunters

For many years scientists looked for bones of the ancient hunters themselves. But every time a skull or skeleton was found, it turned out to be no more than a few hundred years old.

In 1947, while men were digging a drainage ditch near the village of Tepexpan, about 20 miles northeast of Mexico City, they uncovered the skull and other bones of a mammoth. This region, called the

Valley of Mexico, is dry today. But 12,000 years ago the climate was rainy, and the Valley held a large lake. Herds of southern mammoths wandered around the edges, feeding on the plants. No doubt they often bathed in the lake.

Mexican and United States scientists thought this was a good place to look for human remains, so they dug into the old lake bed in several places. In one excavation they found a man's skull and several of his bones. These were lying under an earth layer known to be 10,000 years old. Therefore the remains are older than that.

Men of the lake region surely dared not attack a mammoth herd. But sometimes they came upon an animal stuck in a swamp and managed to kill it. When bones of a mammoth were dug up from the lake bed in 1952, one of the spear points that killed it was still lodged between the ribs. A scraper and three stone knives lay near the bones.

Why did the hunters leave these tools behind? Perhaps, while they were butchering the carcass, the smell of meat attracted a great sabertooth which came and drove them away.

Scientists continue to search for relics of the ancient American hunters. No doubt they will find many campsites, hunting grounds, weapons, and human and animal skeletons. Each discovery will tell us a little more of the wanderings of the great game herds and the people who followed them.

Eskimos Conquer the North

IN the year 1888, the skull of a Great Hunter was dug up from a cave in Chancelade, France. Scientists found it was shaped like an Eskimo's skull. This set some people to thinking that the Great Hunters were ancestors of the Eskimos. In the course of time, bands of Hunters may have wandered across Asia and reached North America.

Nobody can prove this, but it is quite possible. The Eskimos use lamps and harpoons like those of the Great Hunters. Perhaps their ancestors *were* Great Hunters who migrated from southern Europe and took their implements with them to the north. The Eskimos surely inherited many of their skills from the Great Hunters. This ancient knowledge, along with later inventions of their own, enabled them to settle in the Arctic — a region where the Ice Age continues to this day.

Ice Age People Today

The arctic shore is one of the wildest frontiers on earth. Here the sea, rather than the land, makes life possible. The waters are vast meadows of plants. You need a microscope to see these plants, but there are billions of them. Tiny shrimplike animals eat the plants, and fishes eat the shrimplike creatures. The fishes in turn serve as food for seals, walrus, and whales.

The ancestors of the Eskimos reached the arctic seas and learned to live from their riches. When the water

was ice-free, they hunted from boats. When it was frozen over, they hunted from the ice. The sea creatures often migrated, and then the Eskimos had to travel far to find new hunting waters.

Since the seas were icebound most of the year, the Eskimos needed other means of travel than boats. So they used sleds and trained their dogs to pull them. This means of transportation was perfectly suited to the region. During most of the year, a shelf of ice extended like a smooth highway along the shore, and over this highway the Eskimos could travel fast and far. Feeding the dogs was no great problem, for they

ate the same food as the hunters – game taken from the sea.

Today the Eskimos are spread over the whole long stretch of shore from Siberia to Greenland. Many now live in towns and use modern boats, tools, and guns. But some bands still live in the ancient way.

The Caribou Drive

A number of groups make their homes along the bays and straits of northern Canada. They hunt caribou when the herds pass by on their seasonal migrations. The main hunt is in autumn; then the animals are fat and meaty and have a new coat of fur that is just right for garments.

The Eskimos know certain places where the herds must cross rivers and inlets on their way to the south.

They go to such a place and lay a trap across the route. Heaps of stones are built a certain distance from one another, in two rows leading to the water. The rows are placed wide apart at the opening, but gradually come together to make a sort of funnel. On top of each heap the Eskimos place something to scare the caribou. This may be a pair of gull's wings dangling from a line so as to flutter in the wind, or it may be an old coat.

Scouts post themselves on lookout hills to watch for the herds. They wait patiently, hour after hour, day after day. At last they spy, far off on the tundra, something that looks like a moving thicket. But the branches and twigs are the antlers of a great herd of caribou.

When the herd enters the mouth of the funnel, the women and children steal behind them. All at once they leap up, waving pieces of skin and howling like wolves. The deer run toward one side of the passage, but are frightened by the stone heaps and the fluttering objects on top of them. The deer turn back. Taking the only route that seems open, they run down to the water, plunge in, and head for the opposite shore.

The men have been waiting, out of sight, in their skin-covered hunting boats. Now their paddles flash in the sun, as the kayaks dart toward the swimming herd. The hunters take up their lances and thrust them again and again, killing the helpless animals.

A great deal of meat and many skins are obtained quickly, but both may have to last a long time. Most of the meat is kept in pits for storage. The skins are spread out to dry. In winter, when there is little light for outdoor work, the women and girls will busy themselves stitching the caribou skins into warm new garments.

Winter on the Ice

In winter, most of the Eskimos of Canada live on the frozen sea and hunt seals. Large islands shelter the straits and gulfs from ocean currents, so the waters freeze over completely. When the ice is firm, and when there is enough snow on it to build houses, the groups of families pack their belongings on sleds and move out on the ice.

The leaders of a group are careful to locate their settlement behind the shelter of a "pressure ridge," where wind and tide have piled up a rampart of ice, and snow has gathered in drifts behind it. The families set to work making houses, which they build as close together as possible.

How to Build a Snow House

When a man has picked a spot for his house, he takes a long knife and cuts the snow into blocks. Then he sets the blocks in a circle. He cuts additional blocks from the snow inside the circle and in this way digs the floor of the house deeper, while he builds up the wall.

When the wall is shoulder-high, he begins to tilt the blocks inward to form a dome. Finally, only a small hole remains in the top of the dome. The builder, still working from inside, slips a block edgewise through the hole, then eases it down into place, closing the dome.

Having shut himself in, the builder cuts a hole in the wall and crawls out through it. The entrance will be here, so he is careful to locate the hole crosswise to the wind. In this position, snow won't drift over the doorway. Then he cuts a path in the drift leading to the door and builds a covered passageway over it.

Inside the house a portion of the drift opposite the doorway is left higher than the rest, so that it forms a platform. This is the place for the family bed, where grownups and children will sleep side by side, wrapped in their fur sleeping bags.

Children of the Frozen Sea

The lamp that heats the house is a long tray made of soapstone. It rests on a wooden framework placed near the wall, between the bed and the door. The women and girls of the family tend it. They arrange pieces of seal fat along its back edge. The lamp wick is made of fluff from the sedge called "cotton grass." The women dip the fluff in melted blubber and pinch it into a ridge running along the front rim of the lamp. They light this wick, and it burns brightly. The lamp is the family hearth. It brightens their long winter night, cooks their food, and enables them to live comfortably in the cold.

The families connect their houses with covered passages that all branch from one entrance. Sometimes two or three houses are built side by side, with openings in the walls between them, so as to make a two-family or three-family house.

Since the houses form an unbroken chain, a person can get from his own to any other house without stepping out of doors. This is important, for when a blizzard whips across the ice, the people may have to stay indoors for days at a time.

The Eskimos sleep comfortably in their snow houses on a sea of ice. They become used to this sort of home from childhood. When boys and girls are snug in their sleeping bags and their eyes are closing, tired from staring at the lamplight, they hear a low, musical tone. It comes from the sea beneath their bed. The tide is stirring there. It presses the ice, which moves slightly and makes a sound of moaning.

If these boys and girls lived in some other place on the earth, they might fall asleep to the sound of leaves rustling in the wind. But they have never seen a tree, even when they were living on land. Now they are people of the frozen sea, and the sound that lulls them to sleep is the moaning of the ice.

Safe from the Blizzard

When the blizzard comes, all hunting stops. It would be suicide to face this terror. The gale whips the warmth out of a body, no matter how well clad with fur. Ice powder blasts the face, chokes the lungs, smothers a man. Animals of the land burrow in the snow or shrink into a rock shelter. Many die.

The blizzard howls around the ice blocks of the ridge and over the drift behind it. If you could be there, you would see nothing except the ice ridge and

drifted snow. But you might hear a song rising out of the drift. People live there, half buried in their little cluster of houses. They are gay in the shelter of the drift. They laugh at the blizzard.

Hunting Seal

The game hunted from the ice is the ringed seal, so called because it has ring-shaped markings on its back. Fortunately for the Eskimos, the ringed seal does not leave these waters when they freeze over. It has a method of living beneath the ice.

In autumn, when the ice is still thin, the seal swims about freely. It comes up anywhere to breathe, for it can break the ice film with its head. When the ice becomes too thick to ram, the seal has to bite holes in it to get air. The seal keeps a number of holes open, and has to stay near them, feeding on fish in the surrounding water.

As the ice grows thicker, the seal has to make passages through it in order to get up to the breathing holes. By working hard, the seal manages to keep a few passages and breathing holes open. The holes film over with ice, which soon is covered with snow. When the seal comes up to use a hole, it breaks the ice film and breathes through the snow.

The seal hunter takes a dog with him to smell out a breathing hole. When a hole has been found, the hunter stands over it and waits for the seal to come up. He may wait for several hours. If he is lucky and the seal comes up here, the hunter harpoons it through the hole. The animal starts to swim away, but the harpoon point is hooked into its flesh and holds firmly as the hunter pulls the seal back to the hole and kills it with a knife.

Eskimo hunting weapons

The body of the ringed seal is a treasure containing nearly everything the Eskimos need for their winter life. Its flesh feeds them and their dogs. Its hide supplies material for their harpoon lines, their kayaks, and their boots. Its blubber is fuel for their lamps. Without this source of warmth, the Eskimos probably would never have been able to settle the icy arctic shores.

The Polar Eskimos

The northernmost point of the earth inhabited by human beings is the northwest coast of Greenland. Here, over wide stretches, the ice sheet reaches the shore and sends glacier snouts plunging into the sea. Only here and there a fringe of shore is left uncovered. A group of Eskimo families live on such a fringe, between two massive ice fronts. The one to the north of them overruns 60 miles of shore; the one to the south, 200 miles.

The ancestors of these people settled in Greenland about a thousand years ago. The group probably has never numbered more than 250 men, women, and children. They are known as the Polar Eskimos, because they dwell only a few hundred miles from the North Pole.

This is a strange place to choose as a home. The winter darkness lasts two months. The people cannot hunt then, so they must gather supplies of food before the darkness comes. They work hard in spring, when there is no night, hunting long hours with little rest, so that their storage pits will be filled with meat.

Reindeer and musk oxen once roamed this northwest fringe of tundra, but the herds have been wiped out. They were shut into such a small space that the Polar Eskimos easily hunted them down. Now the people get almost all their food from the sea.

Along other stretches of the Greenland coast, to the south, there are towns. Here people make their living as commercial fishermen. But they, like the Polar Eskimos, live near the edge of the ice sheet.

During the past century, the glacier snouts have retreated. At other times they have advanced. Perhaps they will advance again and plow over the entire shore.

If the ice front overwhelms the fringe of tundra around northwest Greenland, the Polar Eskimos may be able to leave by ship. But even if there are no ships, these conquerors of the North no doubt will escape the flood. They have found new homes before. The ice has never defeated them.

Is Another Ice Age Coming?

THE Ice Age has not ended. Glaciers still flow from high mountains in many parts of the world. Icecaps cover several northern islands. All of Antarctica and most of Greenland are buried. It is true that the world's climate has been rather mild during the past few thousand years, but how long will this last? Will ice sheets spread again?

The Shifting Snow Line

Ice gathers on mountain heights above the snow line. There, summers are so cool that only part of the winter's snowfall melts. The rest remains, adding to the snow of other years. Below the snow line, the climate is too warm for snow to accumulate.

In the Alps at the present time, most snow falls 2,000 feet below the snow line, landing in places where it will melt away during the summer. But suppose the average summer temperature were to drop six degrees Fahrenheit. Then the snow line would be lowered 2,000 feet. The heavy snowfall at this level would pile up and make new ice fields and greater glaciers.

Very cold winters have an opposite effect from cold summers. The colder the air, the less moisture it will hold, and the less snow falls. But when winters are warmer, the air is wetter and brings more snow.

Most of northern Europe has fairly warm winters. The mildness is caused by the Gulf Stream, which flows through the ocean from the tropics, brings warm water to the north European coast, and warms up the air. Since the air is warm, it picks up a great deal of moisture from the ocean. Winds bring the moisture to the mountains, where it falls as snow.

If summers were to become just a little cooler and winters just a little warmer, much more snow would pile up in the mountains of Scandinavia. The ice fields would expand, and the glaciers would thicken and lengthen. If this went on for a few thousand years, the glaciers would pour over the lowland and form a new ice sheet.

The Earth's Tilt

You know that the seasons come and go because the earth's axis is tilted. As our planet travels in its

The tilt of the earth's axis causes the seasons.

If the earth were not tilted, there would be no seasons.

If the earth were tilted less, summers would be cooler and winters warmer.

orbit, the northern hemisphere turns toward the sun during part of the journey, receives more sunlight, and has its summer. When the earth reaches the other side of its orbit, the northern hemisphere turns away from the sun and has its winter. The southern hemisphere has opposite seasons.

Suppose the earth were not tilted at all. The drawing on page 83 shows this situation. Any point on the globe would receive the same amount of sunlight at all times of the year. The polar regions would always be cold, the equatorial zone hot, and the temperate zones cool. There would be no seasonal changes anywhere.

Now imagine that the earth's tilt were just a little less than it is. There would be seasons, but they would be less marked. Summers would be cooler, and winters warmer. These conditions would cause more snow to pile up and form ice. Present ice fields and glaciers would expand, and greater ones would develop.

Sometimes the earth does tilt a little less, and at other times a little more. Astronomers know that a change from one extreme to the other and back again takes about 40,000 years. This means that, once in every cycle of 40,000 years, conditions should become just right to start an ice age.

A Yugoslav scientist, Milutin Milankovitch, made a calendar of the last million years showing this cycle and certain other periodic changes in the earth's position. He compared the warm and cold periods of the Ice Age with the cycles and found that they fitted pretty well.

We cannot say that these cycles are the only cause

of the climate swings. The cycles have been going on throughout the earth's history, yet climates have been mild during most of that enormous stretch of time. Before the recent Ice Age, our whole planet was warm for several hundred million years. Why did it suddenly become cold? What drastic event could have started the Ice Age?

Wandering Poles

The Ice Age may have been brought about by a change in the position of the earth's poles. There is strong evidence that the poles were formerly in different places. The evidence comes from the study of magnetism in rocks. As rocks form, some of their particles line up like compass needles, pointing toward the earth's magnetic poles. These poles are near the geographic poles, which are at the ends of the earth's axis of rotation.

If the magnetic poles had always been where they are now, the magnetic particles in all rocks would point north and south. But in many ancient rocks, the particles point in other directions. In some, they even point east and west. When these rocks formed, the magnetic poles clearly were not in their present location. At times they were where the equator is now, and sometimes they were in the middle of the Atlantic or the Pacific. (No doubt the geographic poles were near them in the same areas.)

When the North and South Poles lay in mid-ocean, climates must have been quite different from those we

have today. Winters were very mild. Ocean currents brought warmth from the tropics all the way to the poles, and this prevented any accumulation of ice.

How the Ice Age Began

Several million years ago, the poles reached their present sites in the Arctic Ocean and Antarctica. What effect did their new location have? According to two American scientists, William L. Donn and Maurice Ewing, there must have been a great disturbance of the world's climate. This disturbance resulted from the fact that the poles were now isolated from the large ocean bodies.

The upset began in Antarctica. The interior of that continent, since it was far from the ocean's warming influence, grew very cold. But while the ocean sent little warmth to the land, it did send moisture, carried by the winds. The moisture froze and fell as snow, which piled up and gradually formed ice sheets.

In the North, the Arctic Ocean remained ice-free for a long time. A great deal of moisture evaporated from its surface. The same thing happened over the other oceans. The air was wet, and its moisture provided an abundant snowfall. This started the accumulation of ice in Europe, Asia, Greenland, and North America.

The ice sheets made their own climate. Their bright surface reflected much of the sun's heat back into space. The air over the ice grew colder, and cold winds streamed outward, chilling the waters of the region.

The Arctic Ocean froze over. This stopped evaporation from its surface. At the same time, the northern Atlantic and Pacific Oceans became so cold that evaporation from their surfaces was greatly reduced. All over the North, the air became drier and drier. There was less snow to supply the ice sheets, and they began to melt back around their edges.

The shrinking of the ice helped to warm the climate. Newly uncovered land soon was overgrown with plants and inhabited by animals.

Finally, even the ice of the Arctic Ocean disappeared. Its melting was to produce a strange result — a rebuilding of the ice sheets. For with the Arctic Ocean open again, evaporation from its surface provided more snow. Well-nourished ice sheets grew, advanced, and brought about another glacial period.

And so the cycle went, swinging back and forth

from ice advance to ice retreat, from cold to warmth and back again.

Antarctic Mystery

Which way is the cycle going today? Are we headed for another glacial period, or are we on the way into a period of melting and warmth? For over a century, climates of the world have been warming up a little; and glaciers in the northern hemisphere have been retreating. This trend may continue, but we cannot be sure. Other times of warming have ended after a few centuries.

In 1957-58, scientists of the world cooperated on a great study of our planet. This was the International Geophysical Year – the IGY. One important program of the IGY was investigation of the earth's weather and climate. Since the south polar region was the least known place on earth, several expeditions went to Antarctica. Stations were set up on the ice around the edges of the continent and in the interior, and research teams struggled in the cold to learn the secrets of the frozen land. Since the IGY, an international committee has directed further investigations.

The big question about Antarctica has been: How much ice is there, and is the total increasing or decreasing?

To estimate the quantity of ice, it is necessary to measure the thickness of the sheet in as many places as possible. Measurements are taken mainly by a method called *seismic shooting*. An explosive charge is set off, sending a wave of vibrations down through the ice to the underlying rock. The vibrations bounce back from the rock and are recorded by instruments. The speed of such waves through ice is known, so the time lapse between a shot and return of the wave tells how thick the ice is. Recently, scientists have been making measurements with a new type of *radar*. Radio waves shoot down through the ice and return, and their travel time gives the thickness of the ice.

By taking hundreds of measurements in Antarctica, scientists have found that the ice increases in thickness from the continental margins to the interior, where the surface rises in the form of a broad dome. The greatest thickness measured so far is more than three miles. The average for the whole continent is 7,000 feet – considerably more than a mile.

To get the total volume of ice, the area of the continent is multiplied by 7,000 feet, which gives a fantastic figure – 8,000,000 cubic miles! This amount of ice would fill a space a thousand miles long, a thousand miles wide, and eight miles thick.

Ice Sheets and Sea Level

This vast load amounts to 90 per cent of the world's ice and represents 2 per cent of its water, held in storage. If there is to be any great shrinking or increase of the world's ice, what happens in Antarctica will be all-important.

If the entire ice load of Antarctica were to melt, 7,000,000 cubic miles of water would be added to the oceans. This would raise the sea level by more than 200 feet. But since the weight of the added water

would cause the ocean floor to sink about 70 feet, the actual rise in sea level would be about 130 feet. This is the height of a thirteen-story building.

Such an increase would flood the shores of the continents; it would submerge New York, London, Leningrad, and all the great port cities of the world. Vast stretches of land inhabited by millions of people would be lost beneath the sea. Where would the people go? Possibly to Antarctica and Greenland, newly freed of ice.

Such a disastrous melting of the world's ice would not happen suddenly; it would take thousands or even millions of years. But what about the near future? What may people expect in the next few centuries?

The answer depends mainly on whether the Antarctic ice load is growing or shrinking.

Danger Signs

There are signs that the Antarctic ice was thicker than now. Bare mountaintops that rise one or two thousand feet above the surface have ice-carved shapes. Since the time when these summits were sculptured, the level of the ice has fallen at least one or two thousand feet.

Is the ice still shrinking? To answer this question, scientists need many more measurements. They must know how much ice is discharged into the ocean around the coasts, then compare the rate of loss with the rate at which new layers are added by snowfall.

During the last hundred years, the world has become slightly warmer, and the sea level has risen four and a half inches. In harmony with these changes, glaciers of the northern hemisphere have been retreating. Possibly the Antarctic ice also is shrinking, or may soon begin to shrink. If so, low countries like Holland will face flooding. Then other countries too will be threatened.

Melting in the northern hemisphere might go on until the Arctic Ocean again became free of ice. Then there would be abundant evaporation from its surface, wetter air, and more snow. As a result, new ice sheets would probably form in North America, Europe, and Asia. The northern countries would again be in danger.

Can Man Control the Ice?

If this happens, we can be sure that nations will not sit idly by and let ice sheets overrun homelands. No doubt they will apply science to the problem of controlling climate. If ice accumulates to a threatening degree, scientists may be able to melt it by covering the surface with a dark material, which would absorb more of the sun's heat. Or atomic reactors may be used to melt the ice.

Until now, man has adjusted to the ice. In the future, he may conquer it.

Index